Be a Rock SCIENTIST

• Question • Experiment • Discover

By Alix Wood

Ruby Tuesday Books

Published in 2024 by Ruby Tuesday Books Ltd.

Copyright © 2024 Ruby Tuesday Books Ltd.

All rights reserved. No part of this publication may be reproduced in whole or in part, stored in any retrieval system, or transmitted in any form or by any means, electronic, mechanical, photocopying, recording, or otherwise, without written permission from the publisher.

Editors: Ruth Owen & Mark J. Sachner
Design: Alix Wood
Production: John Lingham

Photo credits:
Alamy: 20B (Ognyan Yosifov), 24 (Universal Images Group North America LLC); Ruby Tuesday Books: 7, 12, 13T & C, 15, 19, 21T, 23; Science Photo Library: 8T (Karsten Schneider), 18B (Dr Juerg Alean), 22C (Dirk Wiersma), 22BR (Mark A. Schneider), 24 (Dirk Wiersma), 26TL (Turtle Rock Scientific); Shutterstock: Cover TL (Wavebreakmedia), Cover TC (Sebastian Janicki), Cover TR (kasidit), Cover BL (Aleksa Stanko/Erwin Weber), Cover BR (ArgenLant/AnyaWhy), 1 (Thonghchai.S), 3 (optimarc), 4T (Thonghchai.S/optimarc), 4B (Attila Jandi), 5TL (Helen Hotson), 5T (yumar salas), 5C & B, 6TL (ixpert), 6TR (Felix Lipov), 8B (Tada Images), 9T (beboy), 9CL (MNStudio), 9CR (Ralf Lehmann), 9BR, 10L (schame), 10R (Sean Pavone), 11, 13B (Filip Fuxa), 14T (Franckpoupart), 14B (Viktor Osipenko/bamgraphy/Asimm Graphics), 16T (AlexussK), 16C (Yasonya), 16B (ikunalmathur), 17T (alex7370), 17B (Ermak Oksana), 18T, 20T (Delpixel), 22T (Yellowj), 22BL (Studio492), 24 (Bjoern Wylezich/IamTK/olpo/sonsart), 25T (MNStudio), 25C (Peter Jozefek), 25B (ArgenLant/vvoe/Mirka Moksha), 26TR, 26BL, 26BR, 27, 28T (Bjoern Wylezich), 28BL (LarysaPol), 28BR (elenaburn), 30L (romeovip_md/Summer 1810/vvoe/Nikolay Zaborskikh), 30BR (nnattalli/vvoe/Helen Hotson), 31L (Byjeng/Maria Surtu/ArgenLant/milart), 31R (Kit Leong/Andrey Kuzmin/Gani Prastowo), 32 (24K-Production); Alix Wood: 6B, 21B.

ISBN 978-1-78856-435-9

Printed in Poland by L&C Printing Group

www.rubytuesdaybooks.com

Contents

Rocks Are Amazing!..4
Our Rocky Home 6
Making New Crust..8
Rock Versus Wind and Water 10
Making Rock from Sediment 12
Making Soil with Sediment....................14
How to Make Sand 16
Making Rock with Heat 18
The Power of Ice......................................20
Precious Metals and Gemstones...........22
Rock Properties 24
Sink, Soak and Scratch Tests! 26
Let's Talk Rocks...28
Glossary 30
Index..32

Rocks Are Amazing!

Have you ever taken a close look at a rock? Some rocks form incredible shapes. Others can be beautiful colours.

Quartz rock

The rainbow mountains in China are made from a rock called sandstone.

The Giant's Causeway in Ireland is made of 40,000 columns of basalt rock.

What is rock? Rock is a hard *natural* material.

Rocks in Our World

Go outside and look for rocks.

You might see big rocks.

Do you see little rocks on the ground?

Where does all that rock come from?

Rocks are made of natural, solid substances called **minerals**. A rock can be made of one type of mineral or a mixture of minerals.

You can see the colours of the different minerals in these rocks.

Which of these is made of rock?

Statue

Chalk

A Stone Age arrowhead

Pathway

To find answers and more information, turn to page 28.

Now, it's time to investigate some amazing rocks and...

...be a **Rock Scientist**.

Our Rocky Home

Where does all the rock on Earth come from?

Our planet is completely covered with a rocky **crust**. Large areas of the crust are covered with oceans.

Earth — Land — Ocean

Mountain range

Thicker areas of crust make the land and high mountain ranges.

Mountain range

Thick crust

Ocean

Land

Thick crust

Crust

We can't always see Earth's rocky crust. It's covered by buildings, roads, pavements, grass and a thick layer of **soil**.

Let's make a model of the Earth.

Earth is made up of different layers.

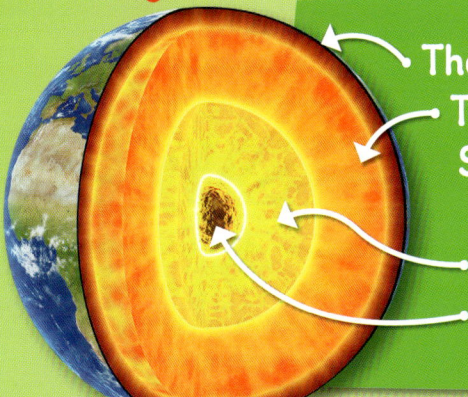

The crust is made of rock.
The mantle is also made of rock. Scientists think in some places the rock is hot and gooey, like melted caramel.
The outer core is made of hot, liquid metal.
The core is made of hot, solid metal.

1) Make four balls of plasticine, each one bigger than the last.

Core   Outer core   Mantle   Crust

You will need:
- Plasticine
- A rolling pin and cutting board
- A plastic knife

2) With the rolling pin, flatten each ball, except the core, into a circle.

3) Wrap the outer core circle around the core. Then wrap the mantle circle around the outer core.

4) Finally, wrap the whole model in the crust.

5) Now add blue oceans to your model Earth. You can also add mountains and green forests to the land.

Cut your model in half with the plastic knife. Can you see and name all the layers?

7

Making New Crust

New rocky crust is forming all the time. How?

Earth's crust isn't one solid piece. It is broken into giant pieces called tectonic plates.

The red and yellow lines show the edges of the plates.

The tectonic plates fit together like a jigsaw.

The jigsaw pieces of crust move very, very slowly.

As they move, their edges rub and crunch. Sometimes, this makes cracks appear.

Crack

Under Earth's crust is hot, melted rock called **magma**.

Sometimes the magma bursts through a crack onto the surface. The place where this happens is called a **volcano**.

Volcano
Magma
Lava

Once magma is on the surface, it is known as **lava**.

Lava
Cooling lava

New hard rock

Dacite

The rock that forms from cooled lava is called **igneous rock**. There are lots of different kinds of igneous rocks.

The lava cools, gets hard and becomes new rock.

This is one of the ways that new rocky crust forms.

Obsidian

Pumice

Rock Versus Wind and Water

Which do you think is toughest — rock, wind or water? Rock is tough stuff. But sometimes, wind and water can be tougher than rock!

When water washes over rock, it breaks off tiny pieces.

The pieces of rock are called **sediment**.

Over many years, water carved this cave in rock.

In a desert, the wind blows sand against rocks. The sand rubs the rocks and makes tiny pieces of rock break off.

The wind made these rock shapes.

It can take hundreds, thousands or millions of years for water and wind to shape rock.

You can see the power of water much faster with this activity!

Do you have super-strong water?

1) Make a stack of sugar cubes.

You will need:
- A box of sugar cubes
- A cup of water
- An eye dropper

2) Fill the dropper with water.

3) Squeeze water from the dropper onto your sugar cube stack.

4) Keep adding water to the same spot.

Does the water change the stack's shape? Describe what happens.

This is a piece of rock called granite.

This is a granite pebble from a river.

How do you think the pebble got its shape? Why is the pebble smooth?

To find answers and more information, turn to page 28.

Making Rock from Sediment

Some rock is made of sediment that breaks off bigger rocks. Here is one way this can happen.

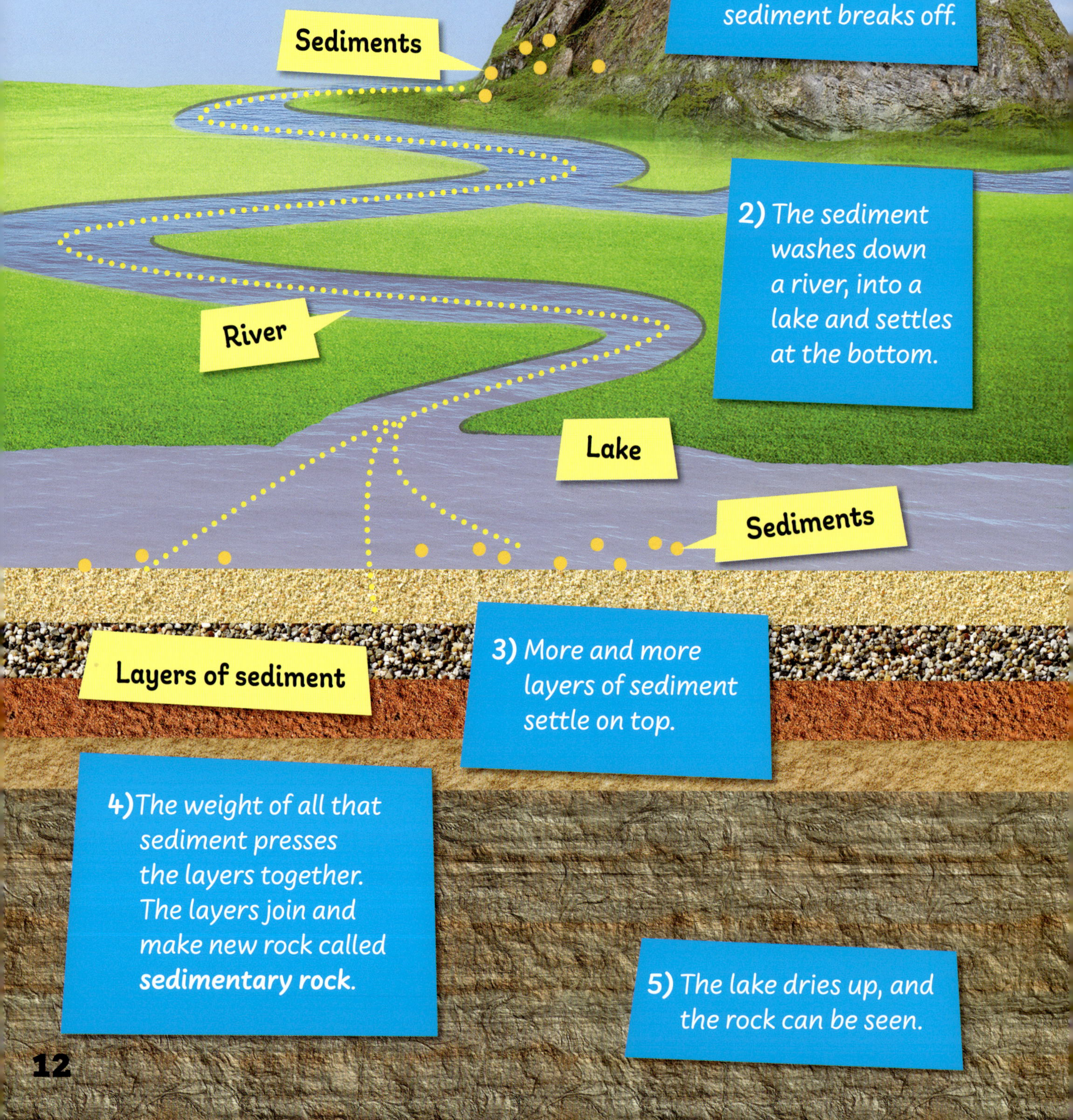

Sediments

1) Rain washes over mountains and sediment breaks off.

River

2) The sediment washes down a river, into a lake and settles at the bottom.

Lake

Sediments

Layers of sediment

3) More and more layers of sediment settle on top.

4) The weight of all that sediment presses the layers together. The layers join and make new rock called **sedimentary rock**.

5) The lake dries up, and the rock can be seen.

12

Discover how sedimentary rock forms by making some chocolate rock!

You will need:
- A measuring cup
- Aluminium foil
- A fine-toothed grater
- An adult helper
- 1 cup each of white, milk and dark chocolate squares

1) Line the cup with foil.

2) Ask an adult to help you carefully grate a small heap of each type of chocolate. This is your sediment.

3) Place a layer of milk chocolate into the foil-lined cup. Press it down with your fingers.

4) Add different-coloured layers of chocolate, pressing down hard on each one.

5) Lift the foil from the cup and unwrap your chocolate rock. (Keep your rock for the activity on page 19.)

Chocolate rock

What happened to your chocolate sediment?

What words would you use to describe your chocolate rock?

Why do you think this sedimentary rock is striped?

To find answers and more information, turn to page 28.

Making Soil with Sediment

Sediment has another important job to do – it makes soil. Most of Earth's rocky land is covered with a layer of soil.

Rocky mountains

When pieces of sediment break off bigger rocks, they become soil.

Soil

Soil also has lots of other ingredients.

Rotting leaves, flowers and other parts of dead plants.

Rain, melted snow and air.

Animal poo and the dead bodies of insects, birds and other animals.

Let's investigate what's in soil!

Soil is different from place to place. That's because it's made of different types of sediment and mixtures of different ingredients.

You will need:
- A small trowel
- Half a cup of soil from a place where plants grow
- A glass jar with a screw-top lid
- Water
- A spoon with a long handle
- A magnifying glass
- A notebook and pencil

1) Tip the soil into the jar. Fill the jar about three-quarters full with water.

2) Tightly screw on the jar's lid. Shake the jar for 30 seconds.

3) Let the jar stand for one hour.

What do you observe has happened?

4) Use the spoon to scoop out any floating ingredients.

5) Scoop out some of the material at the bottom of the jar. Take a closer look with your magnifying glass.

What is at the bottom of the jar?

6) In your notebook, make a list of all the soil ingredients you can observe and identify.

To find answers and more information, turn to page 29.

How to Make Sand

Tiny pieces of sediment make sedimentary rocks and soil. They also make the sand we see on beaches.

A close-up photo of sand

When waves crash against rocky cliffs, tiny grains of rock break off.

Cliffs
Waves
Sandy beach

This rocky sediment is washed up on the shore and makes beaches.

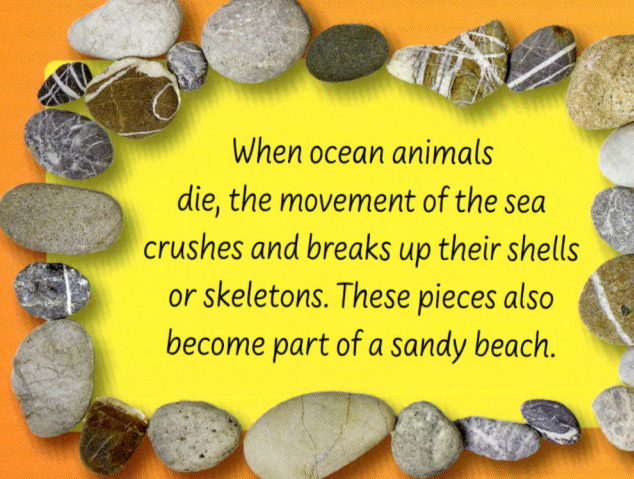

When ocean animals die, the movement of the sea crushes and breaks up their shells or skeletons. These pieces also become part of a sandy beach.

Starfish
Clams
Coral

Be a Sand Detective!

Take a close-up look at some sand and see if you can detect its ingredients.

You will need:
- 3 cups of sand
- A piece of black paper
- A magnifying glass
- A notebook and pencil
- A cup of white vinegar
- An eye dropper

1) Place a handful of sand onto a piece of black paper.

2) With the magnifying glass, take a close look at the sand. In your notebook, make a list of the different things you observe. For example:

- What colours are the grains of rock?
- Do some of the grains look see-through?
- Do you see any pieces of material that look as if they were once part of an ocean animal?

3) With the eye dropper, squeeze some drops of vinegar onto the sand. If the sand starts to fizz and bubble, it means there are grains that were once part of a living thing!

To find answers and more information, turn to page 29.

Making Rock with Heat

Sometimes rocks get recycled into new types of rocks called **metamorphic rocks**.

Deep in Earth's crust, rocks get baked by the heat and change into metamorphic rocks.

As Earth's crust moves, underground rocks get twisted, stretched and crushed.

These movements make heat and **pressure** that change the rocks into metamorphic rocks.

Gneiss

Jaspilite

Soapstone

These metamorphic rocks look very different. That's because they formed from different types of rocks.

You can see twists and folds in this huge metamorphic rock.

Recycle your chocolate sedimentary rock and change it into metamorphic rock!

1) Take a photo of your chocolate sedimentary rock. Tightly wrap the rock in foil to make a parcel.

You will need:
- Your chocolate rock from page 13
- A phone or camera
- Aluminium foil
- 2 pairs of tongs
- An adult helper and a saucepan of hot (but not boiling) water

2) Using tongs, float the parcel on the hot water in the saucepan. Make sure your adult helper is close by.

3) Test if the chocolate is melting by squeezing the parcel with the tongs. If it's soft, remove the parcel from the saucepan.

4) Using tongs, twist the hot parcel. Leave the chocolate to cool and harden.

5) Unwrap your new, recycled chocolate metamorphic rock.

Chocolate metamorphic rock

How has the chocolate rock changed?

How is what happened similar to what happens to real rocks?

If metamorphic rocks form deep underground in Earth's crust, how do you think we find them?

To find answers and more information, turn to page 29.

The Power of Ice

We know that water and wind can shape and change rock. Water can be even more powerful when it turns to ice.

A glacier is a giant mass of ice that's always slowly moving downhill.

Mountains

Bottom of glacier

Valley

Glacier moving downhill

A glacier can carve a huge valley into a mountain range.

This takes thousands of years!

Boulder

A glacier can move rocks and giant boulders. The glacier that moved the boulder in this picture has now melted.

Make a Rock-Moving Glacier

You will need:
- A sandwich bag
- Some small rocks
- Water
- A freezer
- A long tray, at least 2.5 cm deep
- Soil or sand and small rocks
- A plastic bowl

1) Begin by making your glacier! Place some small rocks in a sandwich bag. Fill the bag with water and seal it shut.

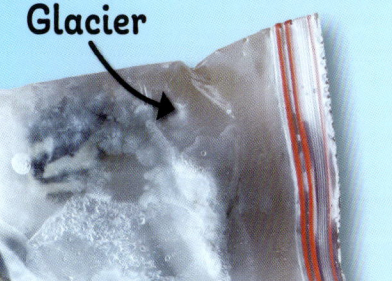

Glacier

2) Lay the bag flat on a shelf in the freezer, and leave it overnight.

3) Cover the tray with soil, sand or a mixture of both. You can also add small rocks.

4) Outdoors or on a waterproof surface, place one end of the tray on the plastic bowl, so the tray is slightly sloping.

5) Take your glacier from the sandwich bag. Place the ice on the tray at the top of the slope.

Watch as the ice glacier slowly melts and slides down your slope.

What happened to the rocks that were frozen in your glacier?

What happened to the ground below your glacier?

Did your glacier carve out a valley?

In real life, as a glacier melts at the bottom of a slope, new ice forms higher up.

Precious Metals and Gemstones

All rocks are made of minerals. But some rocks contain minerals that are so **precious** we use them to make jewellery.

Gold is a mineral that is sometimes found in rocks.

Gold is also a type of metal.

Sparkling diamonds are minerals that form in Earth's crust.

The heat and pressure deep underground make diamonds grow in shapes called **crystals**.

Some minerals are so beautiful they are known as gemstones. Rubies and sapphires are gemstones that form in rocks from a mineral called corundum.

Sapphire, diamond and ruby earrings

Many types of minerals form as crystals.

A crystal is a solid shape with straight edges and smooth sides, called faces.

Quartz crystals

Grow Your Own Crystals

Salt is a type of mineral. Let's grow salt crystals.

You will need:
- A small saucepan
- ½ cup water
- An adult helper
- ½ cup Epsom salts
- A spoon
- Food colouring
- A small glass jar
- A lolly stick
- String
- Scissors

1) Add the water to the saucepan. Ask your adult helper to heat the water until it bubbles.

2) Add the Epsom salts to the hot water. Stir the mixture until the salt dissolves.

3) Add a few drops of food colouring. Then pour the mixture into the jar.

4) Tie a piece of string to the lolly stick. Lay the stick across the jar so the string dangles inside the salty mixture.

The string should not touch the sides or bottom of the jar.

5) Put the jar in the fridge for three hours.

6) Remove the jar from the fridge. Slowly lift up the stick. You will see crystals growing on the string!

23

Rock Properties

Scientists who study rocks are called geologists. When a geologist finds a rock, they examine it to discover its **properties**.

A geologist wants to know how a rock was made. Is the rock igneous, sedimentary or metamorphic?

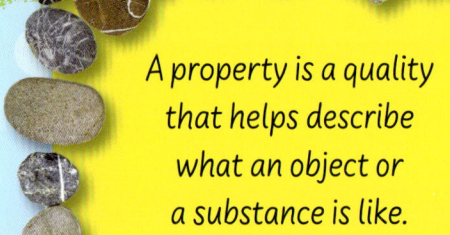

A property is a quality that helps describe what an object or a substance is like.

Snowflake obsidian

Taconite

Serpentinite

Quartzite

Andesite

Conglomerate

Examine these pictures. Can you sort the six rocks into the three different types?

To find answers and more information, turn to page 29.

Be a Geologist

Go rock-hunting in your garden, school playground or on a beach.

You can also buy rocks online or from a rock collector's shop.

You will need:
- Some rocks
- A notebook and pencils
- A ruler or tape measure
- A camera or phone
- A magnifying glass

1) Choose a rock. Draw your rock or take a photo, print it out and stick it in your notebook.

2) Measure and record the size of your rock.

3) Now describe your rock's properties. These questions and the rock words below will help you.

- Is the rock one colour or a mixture of colours?
- What is the rock's shape?
- How does the rock feel?
- Can you see light through your rock?
- Can you write with your rock on a pavement?

Look at the rock under a magnifying glass.

Can you see tiny grains, crystals, holes or a stripy or spotty pattern?

Rock Property Words

rough	shiny	cold	shimmery
smooth	bumpy	warm	see-through
dull	jagged	crumbly	polished

25

Sink, Soak and Scratch Tests!

You will need:
- Your rocks
- A notebook and pencil
- A small jug of water
- Kitchen scales
- A small jar with an airtight lid

Some rocks actually float. Let's test it!

1) Examine each of your rocks from page 25. Do you think they will sink or float?

2) Write your predictions in your notebook.

3) Place each rock in the jug of water. What happens? Did your results match your predictions?

Some igneous rock, such as pumice, has lots of holes in it. Air fills the holes and makes the rock float.

Some rocks can absorb, or soak up, water. Let's test it!

1) Examine your rocks. Choose the one you think is most likely to soak up water. Weigh the rock and record its weight.

2) Pour water into a jar. Put the rock in the jar and tightly close the lid. Leave the rock in water overnight.

3) Take the rock from the water. Shake off any drips. Weigh your rock.

Why did you choose that rock?

Is your rock heavier?

What do your results tell you?

Is a rock hard or soft?

A rock scientist scratches a rock with different objects. Then they give the rock a number from 1 to 10.

Talc — 1 is the softest.

Diamond — 10 is the hardest.

You will need:
- Your rocks
- A copper coin
- A stainless steel nail
- A piece of quartz
- A soft cloth

You are going to scratch your rocks with different objects. The objects' hardness numbers are below.

Fingernail
2.5

Coin
3.5

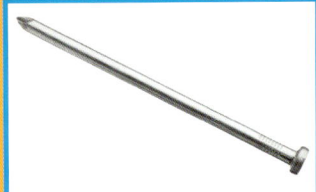
Stainless steel nail
5.5

Quartz
7

1) Scratch a rock with your fingernail. Wipe the rock. Did the scratch leave a mark?

2) If not, try scratching with the coin, then the nail, then the quartz, in that order. Stop when one leaves a mark.

If you can scratch your rock with the stainless steel nail but not the coin, then your rock's hardness is between 3.5 and 5.5 out of 10.

Put your rocks in order from softest to hardest.

3.5

5.5

6

more than 7

Let's Talk Rocks

Did you enjoy being a rock scientist? Let's check out some answers and discover more cool things about rocks.

Page 5:
Which of the pictures is rock? They all are!
The statue was carved from a rock called marble. Chalk is soft, but it's actually a type of rock. The arrowhead is about 7000 years old. It was carved from a rock called flint by a Stone Age hunter. The pathway is made of blocks of granite rock.

Page 11:
Did the water change the shape of your sugar cube mountain? This is what happens when rainwater or a stream or river flows over rock.

When water, snow, ice or wind breaks tiny pieces off a rock, it's called weathering.

The granite river pebble once looked like the rough piece of granite rock. The river water washed over the rock, breaking off pieces. The water also rubbed the rock against other rocks. Over time, this made the rough rock become round and smooth.

Weathered beach pebbles

Page 13:
The pressure of your fingers pressing on the chocolate layers made them join together. This is a little like what happens when heavy layers of sediments press together and make rock.

Sedimentary rock often has striped layers. Each layer shows us a type of rock that was once in the area and became sediment. It can take thousands or millions of years for sedimentary rock to form.

A sedimentary rock called banded ironstone.

Page 15:

The soil in your jar probably separated into different ingredients. The heavier rock sediments were at the bottom. Did you see twigs or tiny pieces of plants floating in the water?

Floating pieces of leaves and twigs.

Sediment

Soil might be soft and crumbly with lots of rotted material.

Soil might be dry and hard with lots of rocky pieces.

Page 17:

The shells and skeletons of living things contain a mineral called calcium carbonate. When vinegar touches calcium carbonate, it fizzes and bubbles. This is because a chemical reaction happens and makes a gas called carbon dioxide. It's carbon dioxide that makes fizzy drinks bubbly!

Page 19:

Metamorphic rocks form deep in Earth's crust. As wind and rain wear away Earth's surface, the top layers of rock crumble and are blown or washed away. Over millions of years, rocks that formed underground become Earth's new surface layer.

When movements happen in Earth's crust, metamorphic rocks from deep underground get pushed up to the surface.

Page 24:

Did you sort the rocks into the correct types?

Igneous

 Snowflake obsidian

 Andesite

Sedimentary

Taconite Conglomerate

Metamorphic

 Quartzite

 Serpentinite

Glossary

crust
The outside layer of Earth that's made of rock.

crystal
A solid substance that has formed a shape, or shapes, with straight edges and flat, smooth sides, called faces.

igneous rock
Rock formed from lava on Earth's surface. Also, igneous rock, such as pegmatite, which forms inside Earth's crust when magma cools and hardens.

Pegmatite

lava
Hot, liquid rock that has escaped from inside Earth. When lava is inside Earth, it's called magma.

Magma
Lava

magma
Rock that has melted and turned into a thick, super-hot liquid by heat inside the Earth.

metamorphic rock
Rock that has changed from one type to another because of extreme heat or pressure.

mineral
A solid substance found in nature that makes up rocks. It's possible to see the different minerals in some types of rocks.

natural
Something that is made by nature, not by humans. For example, rocks, grass and clouds are all things made by nature.

precious
Very special, rare, beautiful or valuable.

pressure
A strong push or press from above, below or the side. For example, heavy rock layers press down, or put pressure on, the rocks below them.

property
A quality that helps describe what an object or substance is like. For example, a rock's properties may include being hard and smooth.

sediment
Tiny pieces of rock that have broken away from larger rocks.

sedimentary rock
Rock made from many layers of sediment that have been pressed together. The layers join and become rock.

soil
A layer of black or brown crumbly material that covers some land. Soil is made from tiny grains of rock and rotted material such as dead plants and animal poo.

volcano
An opening in Earth's crust that allows magma to escape and become lava. As lava cools and turns to rock, it can form a mountain shape.

Index

C
crust (Earth's) 6–7, 8–9, 18–19, 22, 29
crystals 22–23, 25

G
gemstones 22
geologists 24–25
glaciers 20–21

I
igneous rocks 9, 24, 26, 29

L
lava 9
layers (Earth's) 6–7

M
magma 9
metals 7, 22
metamorphic rocks 18–19, 24, 29
minerals 5, 22–23, 29

P
properties (of rocks) 24–25, 26–27

S
sand 10, 16–17
sedimentary rocks 12–13, 16, 19, 24, 28–29
sediments 10, 12–13, 14–15, 16, 28–29
soil 6, 14–15, 16, 29

V
volcanoes 9

Some other planets are rocky, too. Our closest planet neighbours are Mars, Venus and Mercury. They also each have a rocky crust.

The rocky surface of Mars